From the Ground Up...

It would be impossible to write the history of Arizona without including a long and colorful chapter on mining.

From the time Coronado came searching for the Seven Cities of Cibola, the quest for mineral wealth has never died, and the history of Arizona is etched with tales of gold, silver and copper.

More than 400,000 mining claims have been recorded in Arizona, and it is estimated that over the years more than 4,000 companies were formed for the purpose of mining.

In 1981 Phelps Dodge Corporation, then observing its centennial year in Arizona, commissioned former Governor Jack Williams to research and write these stories about Arizona mines and early mineral discoveries.

These are tales told and re-told around campfires by early prospectors, pioneers, soldiers and mountain men...and as with most stories about mineral discoveries, there are several versions for each. Governor Williams has selected the most commonly agreed-upon versions.

Jack Williams has been an Arizona resident since his childhood. He is a former radio broadcaster and newspaper columnist. He was mayor of Phoenix from 1955 to 1960 and governor of Arizona from 1967 to 1975.

Governor Williams is an avid student of Arizona, its color, romance and history. In "From the Ground Up," he recounts a few of his favorite mining yarns from the state's past, and captures the flavor of an early time when the state was still a rough-and-tumble territory and mining was just getting started.

Today about one of every eight jobs in the state depends on the copper mines. And tax revenues generated directly and indirectly by Arizona's copper industry amount to about one of every four dollars paid to state and local governments.

This industry that is so vital to Arizona started humbly, but with daring, investment and hard work it grew...from the ground up.

"From the Ground Up" was originally published in 1981, but the excitement of Arizona mining continues. In this second edition, printed in 1993, Ken Phillips, Chief Engineer for the Arizona Department of Mines and Mineral Resources, has added an epilogue to each of Governor Williams' stories.

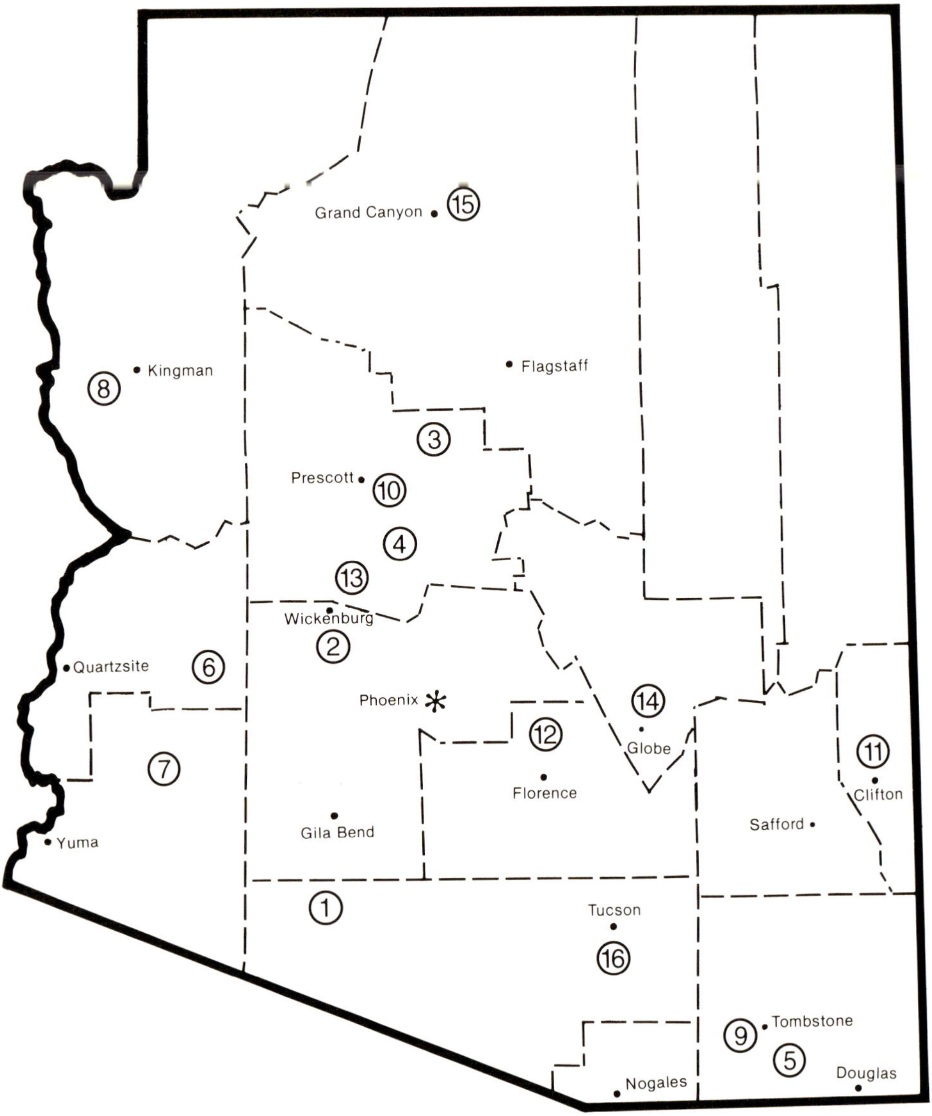
Grand Canyon
15
Kingman
8
Flagstaff
3
Prescott
10
4
13
Wickenburg
2
Quartzsite
6
Phoenix
14
12
Globe
7
11
Clifton
Florence
Safford
Yuma
Gila Bend
1
Tucson
16
Tombstone
9
5
Douglas
Nogales

The Stories...

Investors, Inventors, Ingenious Engineers

New Cornelia Mine. Ajo, Pima County.

Not all bonanzas were discovered by prospectors chasing wandering burros, by blind luck, or by prospectors' picks. Some required a large cast of strange characters.

Although mineralization was known in the area as early as 1750, it was around the beginning of the 20th century that A.J. Shotwell used three low hills in the center of Ajo's basin to attract his gullible investors.

One investor was John R. Boddie, a St. Louis dry goods salesman. He was attracted to the hills, which were stained by copper carbonate assaying one or two percent copper.

Shotwell laid his plans well. He gophered after rich streaks of ore, stockpiled it and later showed possible investors how they could dip their knife blades into a solution of copper carbonate and weak sulphuric acid. The steel was at once coated with pure copper!

The scam was on! Boddie sold interests to all his friends in a mining company named for his first wife—The Cornelia Copper Company.

The story is a long one and includes promoter Shotwell being duped himself, by a forty-year-old medium who brought him in touch with a beautiful spirit maiden, "Little Bright Eyes."

Her seances produced remarkable information, including

directions on how to tunnel. Every turn meant twenty dollars for the medium, so before long the tunnel looked like a flattened corkscrew! In time it made a perfect circle and the medium lost a customer.

The gullible Boddie next engaged the inventor of the McGahan Vacuum Smelter, which supposedly was constructed to melt off all the elements of the ore—gold at the bottom, then silver, copper, calcium, sodium and silicon. Better yet, once started, the smelter ran without any fuel—it ran forever!

Shortly before the test of the smelter, the inventor vanished and the awful truth burst upon Boddie and his friends. They had been fools and dupes.

Finally, when all seemed lost, science replaced the charlatans and the ignorant promoter.

John C. Greenway, who had just finished successfully mining and milling low-grade iron ore at a big profit, decided to drill the Ajo hills. Bad starts didn't discourage Greenway. Within two years, 25,000 feet of drilling proved that those Ajo hills were underlain by 30 million tons of 1.5 percent copper ore.

But the problems of treating such ore had not been solved, and Dr. L.D. Ricketts was called upon to develop a method. A pilot plant was built at Douglas and ran for several months, treating one ton of ore a day.

Water was needed to treat the ore at Ajo, and a deep well was drilled, producing an inexhaustible supply. In the end all worked out—the ore yielded exactly the estimated grade; the leaching plant generated the estimated recovery; and the cost remained within the estimate—a magnificent triumph of engineering skill.

Ajo is still in operation today as the New Cornelia Branch of Phelps Dodge Corporation.

Phelps Dodge's New Cornelia Mine at Ajo currently is on standby status. The slag produced by the New Cornelia smelter is now processed into a variety of industrial mineral products such as roofing granules and sand blasting media.

A Man, a Burro & a Vulture

Vulture Mine. Wickenburg, Maricopa County.

Henry Wickenburg followed the lure of gold from his native Austria to the wild West of the United States. For ten years he searched for the elusive vein that would make him rich. Somehow it was always just beyond his reach.

Wickenburg missed the chance of a lifetime when he learned that the celebrated Peeples-Weaver party had just embarked for Arizona. Wickenburg raced after them and did indeed catch up—just after they hit their great strikes at Weaver's Gulch and Rich Hill. The "big one" had eluded him again.

Discouraged, the prospector and an unnamed companion took to the road. The companion became ill and was forced to return to camp, leaving Wickenburg to continue the search. He was alone...a stranger in these parts, thousands of miles from home...with ten years of searching, always hoping, never succeeding.

Now, to top things off, his burro was growing balky. Wickenburg had a desert to cross, but the animal was moving more and more slowly...slowly...until finally it refused to budge another step. A circling vulture landed nearby, eyeing the stubborn burro. Frustrated beyond endurance, Wickenburg picked up a rock and hurled it at the vulture in disgust.

The rock split open. Wickenburg had discovered gold!

The first gleaming nugget was traced to a rich lode nearly 15 feet wide. Wickenburg's partner put no faith in the prospector's story, so Wickenburg returned to develop the claim himself.

The mine, naturally named the "Vulture," became one of the richest producers in the area. The only drawback was the absence of water nearby, so the ore had to be washed in the Hassayampa River 12 miles away. Many men were killed by Apaches during their trek to and from the river.

With grinding devices constructed, Wickenburg went into the business of selling ore on a cash-and-carry basis. For $15 a ton, a man could mine his own, transport it to the river and mill it for whatever he could get. It was a popular proposition, and men flocked in from miles around to take their chances.

The ensuing rush continued until all the easily accessible, high-grade ore was taken. At this point Wickenburg sold a 4/5ths share of the mine for $75,000 to a Philadelphian, who built a 40-stamp mill nearby.

The Vulture Mining Company operated steadily from 1866 till 1872. During this time it produced $2.5 million in gold from 118,000 tons of ore, not including the gold smuggled out by "high-grader" miners, who loaded their pockets and lunch pails with whatever they could carry home.

The rich lode petered out, but the large quantities of low-grade ore which had been passed over earlier became profitable. The ore was in such demand that buildings constructed of the old mine-dump rock were torn down to be milled for whatever they might contain. Then there was nothing left to mine, and the Vulture faded into history.

Named for the Austrian prospector whose stubborn burro attracted the vulture, the town of Wickenburg still flourishes some 53 miles northwest of Phoenix.

Old mines die hard, if ever! Higher gold prices and modern technology breathed new life into the Vulture Mine in the late 1980s. Tailings from previous extraction plants were agglomerated, then heap-leached to recover the gold. Exploration has identified a near-surface resource that may be mined in the future. Many of the historic buildings are intact, and caretakers living at the Vulture Mine, currently conduct tours of the property or allow tourists to take self-guided tours for a fee.

In the Name of God & the King

United Verde Mine. Jerome, Yavapai County.

It was on the eighth day of May, in the year of Our Lord 1583, that Antonio de Espejo, Captain by Commission from the Viceroy of New Spain, proclaimed, "It is a mine of great richness. We will take possession."

At his order, two small crosses were constructed of poles and thong. One was set in the stope, and the other near the tunnel.

"Hernandez, Barreto, de Luna—your arms!" The three men made ready their harquebuses. "In the name of God and the King," the Captain proclaimed. Briefly he prayed.

"Make ready. Fire!" Flame and smoke belched from the guns. The guide cried out and bounded away; the Indians who had followed the cavalcade scattered like quail.

This occurred 37 years before the Pilgrims landed at Plymouth Rock and 193 years before the Declaration of Independence.

Captain Espejo led his expedition into Arizona in 1582, traversing wild mountains and steep canyons high above a wide valley with an eastern wall of red sandstone carved by the elements into fantastic designs.

"Magńifico!" the Captain must have breathed. Thus was discovered by men of Spain the first evidence of what later became known as Cleopatra Hill and ultimately the site of Jerome.

When American prospectors arrived in 1876, traces of the old workings on the outcroppings were still visible. In one of the ancient stopes were found stone implements, and to the surprise of the prospectors who knew nothing of the Spanish visitations, a crumbling wooden cross.

Espejo's party was led by a breech-clouted young Indian, proud that he had been chosen to guide the white gods from another world. The party included five horsemen dressed in the worn garb of Spanish soldiers, trailed by a pack mule and a ragged line of native Indians, whose curiosity about the visitors was stronger than their fear.

As the party climbed upward, tall and fragrant pines replaced the juniper, mesquite and cactus of the lower elevations. The guide paused and pointed to a pile of loose rock which had spilled over the slope.

"It is the mine," announced the Captain.

"But small sign of riches here," declared the older of the Spaniards.

When a taper was lighted, the Captain ordered his aide into the tunnel where he broke several pieces of the glistening ore from the wall.

Three centuries later, Al Sieber stopped long enough to stake a claim. He named it "Verde" because of the green stain that first had attracted his attention.

Senator W.A. Clark of Montana eventually developed the United Verde Mine, and made $60 million from it. He built frame houses for the miners with families, and an immense stone dormitory that held a thousand men. Jerome had everything, including fourteen saloons—but no water. At one point the townspeople granted a water-hauling contract to a future Mexican revolutionary, Pancho Villa!

Because of the water shortage, fires plagued both the town and the mine. Three times the people had to camp out on the hillside after their homes burned to the ground.

"Rawhide" Jimmy Douglas (the son of Dr. James Douglas) developed a faulted portion of the United Verde ore body, much to Senator Clark's dismay. Douglas formed a company, called the United Verde Extension, which eventually made a profit of $42 million.

Both the United Verde and the United Verde Extension produced well into the 1930s. The United Verde was sold to Phelps Dodge in 1935 and the UVX ore gave out in 1938. Phelps Dodge mined the United Verde until 1953. During the 70 years of mining in Jerome, the big bonanzas produced almost $600 million in copper, silver and gold.

As Espejo said, it was "a mine of great richness."

Convinced that such "a mine of great richness" has yet to yield all its riches, many companies have applied modern mineral exploration techniques to the United Verde and the UVX mines. A.F. Budge Mining produced ore from a high grade gold zone in the UVX mine in the late 1980s and early 1990s. In 1993 Phelps Dodge continued to explore for copper and zinc at the United Verde Mine.

Fool's Gold in the Bradshaws

Crown King area. Yavapai County.

Fort Whipple's commanding officer didn't take kindly to the idea of his men spending their spare time on mining claims. If they had signed up for a tour of duty in the U.S. Army, they should be drilling, marching, working around the barracks...not rushing off to dig for gold.

But the law said any claim on the public domain had to be worked to be held. So when a soldier had a claim, he would hire a local prospector to do "assessment" work, thereby satisfying the law.

One such prospector saw no reason to exert any effort. He figured that most of these claims were worthless, anyway. So when he got an order for some assessment work on a claim near Prescott, he decided to pass it on to somebody else for a smaller fee, and pocket the difference.

He wandered over to the Tiger mining camp and found two bullwhackers. "Say," he proposed, "you fellas be interested in working on a claim?"

"Yeah, sure," the men agreed. "Just give us the tools."

Armed with drills and shovels, they started digging. The dirt flew steadily until they reached solid rock. At that point, one partner turned to the other and said, "I don't know how to drill; looks like you'll have to drill the hole."

"I thought you knew how to drill!" the astonished bullwhacker replied. "All I know how to do is dig!"

The men found a softer spot and started digging again. This time they uncovered a vein about four inches wide, full of sparkling ore that they identified as iron pyrite, or "fool's gold."

Figuring they had done their job by uncovering anything at all, they returned to the mining camp. The prospectors around the campfire greeted them jovially. "Hey, you find anything up there?"

"Nah...just a bunch of them 'iron pirates.'See...?" A nugget was thrown across the campfire, glinting in its lighted arc. The minor who caught it stared incredulously. "You found this?"

"Yeah...we knew it was a waste of time diggin' up there."

"Man, this is gold—all gold!"

The Crown King area yielded gold strike after gold strike, and Bradshaw City, nearby, grew to a rollicking camp of almost 5,000 people.

Over $1.5 million in gold was produced before 1901, and some silver, copper, lead and zinc were mined as well. The developers built a railroad that connected the Crown King with Prescott. Ore was transported along a winding track until 1927, when the railroad was replaced by a highway.

In 1948 production at the Crown King was stopped. But during its rich years, the mine yielded some $2 million of ore—gold, silver, copper, lead, zinc and, if the two bullwhackers are to be believed, maybe even a few worthless nuggets of "iron pirate."

In the 1980s the dumps and tailings at the Crown King Mine were reprocessed to yield more gold and silver. The Gladiator Mine was refurbished, a mill erected and newly proven ore was mined and processed. The DeSoto Mine yielded new copper ore treated by leaching and precipitation.

The Copper Queen, the Irish Mag & Dr. Douglas

The Bisbee Mines. Cochise County.

The story of Bisbee is fascinating. Its cast of characters is fabulous: Warren, Zeckendorf, Steinfeld, Reilly, Daley, the Williams brothers, the Hoatsons...and above all, the towering figure of Dr. James Douglas.

It all goes back to 1877, when an Army scout from Ft. Huachuca noticed some mineral outcroppings. He didn't have time for mining, so he grubstaked a Tombstone miner named George Warren.

George found copper, but soon lost it. After one of his frequent binges, he bet what became the Copper Queen mining claim on a Fourth of July footrace between himself and a horse with rider. He lost. The claim's eventual worth was in excess of $40 million! He then sold the rest of his property for $925 and proceeded to drink himself into oblivion.

Later, Professor Silliman, a shareholder in the Chemical Copper Company, asked the manager of the company's Phoenixville, Pa., operation to evaluate the Copper Queen Mine. There were problems aplenty, particularly with the smelting of the ore. But this Eastern expert had had many successes in solving such problems. Thus Dr. James Douglas first appeared on the Bisbee scene in 1881.

Douglas was enchanted by the possibilities of the area. Since the Atlanta claim adjacent to the Copper Queen looked promising, Douglas later recommended its purchase to Phelps, Dodge & Co. Douglas was retained by Phelps, Dodge & Co. and started development in May of 1881.

Phelps Dodge later bought the Copper Queen, added a number of small claims and, in 1885, began large-scale operations under the name of the Copper Queen Consolidated Mining Company with Dr. Douglas as president.

The complexity of Bisbee's claims was staggering. Space does not permit the entire story to be told, but two colorful episodes can be related as examples.

The "Irish Mag," a claim named by cantankerous Jim Daley for one of the "ladies" at the far end of Brewery Gulch, was near the Copper Queen, and Daley felt the owners should buy it.

Turned down, he became resentful and beat a trespassing Mexican mercilessly. A deputy sheriff tried to interfere and was shot in the foot. Then, Sheriff Bill Lowther, serving the assault warrant, was fatally shot by Daley. Daley was seen running up Sacramento Mountain behind his house, and that was the last authentic knowledge the world has had of him. Daley left behind his Irish Mag for heirs to squabble over through a decade of litigation.

There was ore all around the Irish Mag, but apparently little of it showed on the surface. "Cap'n" Jim Hoatson, a mine foreman on vacation from Michigan, prowled curiously around the area. Exhausted from the hot sun, he dozed under a mesquite. He dreamed of a great ore body nine hundred feet beneath him. He saw it as plainly as the mine maps he could read so well.

The ore body he envisioned followed layers of limestone. Surface outcrops showed it to be hundreds of feet wide and eighty feet thick.

Entranced by his dream, Hoatson returned to Michigan to interest his bosses and friends in the Irish Mag. They believed in Hoatson's tale and purchased the mine.

Nine hundred feet deep, the ore proved tough to reach and required more money than Hoatson had dreamed of. But it finally worked out and the effort was worth it. The Calumet and Arizona Mining Company was formed and the mine eventually paid $15 million in dividends.

Thanks to Dr. Douglas, who got all the property owners in the area together, peace descended on Bisbee. Phelps Dodge gradually acquired all the claims and operated there successfully until the mid-1970s when profitable ore grades finally disappeared.

The Queen Mine tour now takes visitors on an underground trip back into what was once the queen of the copper mines. Visitors are outfitted with hardhat, slicker, and lamp to see what undergound hardrock mining was all about. The Copper Queen Branch of Phelps Dodge continues to produce copper by leaching the dumps and precipitating the copper with scrap iron.

Ed & the Dirty Jackass

Harquahala Mine. La Paz County.

Ed and his burro, Millie, got along pretty well. Ed knew when Millie was tired and wanted water. Millie knew when Ed was discouraged and needed to be nuzzled a little. Together they traveled the West, looking for gold.

One summer they settled in a lush area atop the Harquahala Mountains. Ed had a small stash of gold that kept them in beans and bacon for several weeks. When the pouch began to flatten, he figured on taking a job on the railroad for a while.

Then three government scientists came up the trail to the mountain top. "We'd like to set up an astronomy observatory here," they explained.

"Well, sure," Ed agreed, a little bewildered. "Only I ain't got a lot of grub here. I was just thinkin' about gettin' a job on the railroad."

"Tell you what. We'll share your mountain and you share our grub, how's that?"

That was fine with Ed. The scientists brought up their equipment, and sometimes Ed and the youngest man would take an afternoon off and amble around the mountain, talking about geology. Ed had as much practical experience as the scientist had book-learning. They enjoyed each other's company.

When the men were called back to Washington, Ed and his friend took one last walk. "I guess me and Millie'll do some more prospecting when you fellas leave," Ed remarked.

"Not up here," the young man said. "You'd be better off trying someplace like...see...over there. That shiny spot on the other side of the canyon? That's where you might find gold."

"Yeah? Well, if I do it'll be half yours."

"No, you keep it. Call it payment for your and Millie's hospitality."

So a few days later, Ed and Millie set off to explore the spot that had been lit up for a brief moment. Not once had the setting sun illuminated the spot again. But he had the location firmly fixed in his mind. There was gold there, all right. Ed posted his claim, loaded some ore onto Millie, and set off to spend the winter in town.

Realizing he wasn't getting any younger, he took in three partners who could do the actual mining while he collected his share of the profits. In the spring they set out for the Harquahalas again, full of plans. But on the way up, one of the men made an unforgiveable mistake. He called Millie a "dirty jackass."

Ed ordered his partners to get lost. No one was going to insult his burro and get away with it. He'd mine the ore himself, he would, and they could forget it! Stunned, the men trudged back to town and reported that Ed thought more of that burro than he did of working a gold mine.

And no one ever saw Ed or Millie again.

While the story of its discovery may be legend, the Harquahala Mine is fact. It produced an estimated $1.6 million between 1891 and 1894, and continued to turn a profit until 1908 when operating costs drew even with the value of the gold mined. Today, the Harquahala Mine is known chiefly for the tale of its discoverers—the crusty prospector and his beloved burro.

Miners have never given up on the second richest mine in Yuma County (Yuma County was split into La Paz and Yuma County in the early 1980s). Many companies have invested considerable time and money into finding more ore at the Harquahala Mine. So far production has come from reprocessing of dumps and tailings.

Charley & the King

King of Arizona (Kofa). Yuma County.

"You go up on the mountain toward the water tanks," Charley Eichelberger said to his partner. "I'll try down in the canyon. We'll meet again in two days. If neither of us finds water, we can head for Squaw Peak and try there."

Charlie Carmen agreed. Eichelberger soon found a water tank where he filled his canteen and sat down under a lone desert tree. He rested, drinking the lukewarm water thirstily, and looked out over the canyon.

Suddenly his experienced eyes caught an unexpected flash of light. It seemed to be coming from a cave some twenty feet away—a cave that had probably been used by long-ago Indians, judging from its soot-blackened walls.

With his prospector's curiosity fully aroused, but afraid to let his hopes run wild, Eichelberger walked slowly into the cave where he had seen that arresting glint.

Sure enough, it was gold!

When Charlie Carmen returned days later, he found Eichelberger in the canyon, glorying in his luck. Water and gold—what more could any prospector ask?

Henry Gleason, the Yuma engineer who had grubstaked Eichelberger in exchange for a share of any findings, wasn't so sure. This might be a fraud. He asked Charley to bring back samples of the ore "so we can see what the assayer has to say about it."

"It's gold, Henry. Trust me," Eichelberger insisted.

"Show me the ore," Gleason replied.

So Eichelberger agreed to bring back several burros laden with ore. The thought struck some town loafers—why not follow him and see where he had staked his claim?

Expecting this, Charley set out early in the morning to elude his followers. But two rose before dawn and tracked him to his canyon. Hiding until they were sure it was safe, they staked claims for themselves. Eichelberger had to go to court to have his prior claims declared valid.

Meanwhile, Gleason was satisfied with the samples of ore, which proved to be worth about $500 a ton. The men took in a third partner, Epes Randolph, who managed the Southern

Pacific Railroad in Arizona. Together they organized the King of Arizona Mining Company and set up a mill.

The ore washing through that mill was so rich that the miners had to shut down the mill every few hours to clean the clinging gold dust from the plates. The mill produced $1200 worth of gold from the first two tons of ore.

A mining town called K of A, or Kofa, sprang up almost overnight. Most of the miners were Cornishmen, dubbed "Cousin Jacks," who brought their families with them. Because of this there were few of the street fights and barroom brawls common to other mining camps.

In its lifetime the King of Arizona Mine had a total production of $3.5 million. The mine flourished from 1896 until 1910, when the rich ore gave out. Kofa became a ghost town. Charley Eichelberger spent the rest of his fortune and his life trying to revitalize another mine near Quartzsite, without success. Apparently one gold strike in a lifetime was enough.

Fueled by a story that underground mining was done only if the exposed working face assayed profitable ore, the King of Arizona Mine continues to interest exploration companies.

Over 800,000 tons of tailings remain from processing the ore. As no process ever recovers 100 percent of the gold, the King of Arizona tailings continues to attract metallurgists' interest in extracting some of the gold left by the turn-of-the-century processing. Thus far there have been many attempts and no successes.

Jose, Tonto & the Grocery Bill

The Gold Road Mine. Mohave County.

Many "bonanzas" in the early days were found by gamblers, naive prospectors, stock swindlers, and wanderers who just lucked out. Only those present at the time actually knew how the discoveries really were made, and all those participants and witnesses are long gone.

Jose Jerise, his Indian wife, Maria, and their nine children lived from day to day. During hard times Jose would ask Henry Lovin, a storekeeper at Kingman, to grubstake him to tide the family over.

No one knew exactly how the Jerise family kept alive; but survive they did, trapping rabbits, cultivating a few crops during the rainy months, and depending on the generous credit of Señor Lovin.

Always Jose kept hinting about the rich ore he had seen somewhere, but he couldn't quite remember where. Always he promised to bring back a few samples and lead Lovin to his find.

At long last, in 1899, Jose's bill crept up to the astronomical height of $13.85, and Lovin called it quits.

"Either you find that claim, or no more grubstake from me!" Sadly, Jose went back to his little home, his hungry children and his patient Maria.

"Que lástima!" he sighed. "What can I do? There is no claim and Señor Lovin now knows it! But, I did promise to find it and that I must begin to do."

Jose decided to start looking for a gold mine the very next day. Early the next morning, he found Tonto, his little burro, feeding near the shack. He placed himself firmly on Tonto's back and gave the creature its head.

Tonto plodded off, turned down the road and, with no guidance from the rider, paused now and then to graze.

Finally, Jose and his burro reached an arroyo near Sitgreaves Pass where red-stained cliffs sometimes held back swirling flood waters, but which on this day provided only cooling shadows.

Feeling no urge to continue, Tonto paused and Jose slid off to nap in the shade. When Jose awakened Tonto was nowhere to be found. While searching for his burro he picked up a few interesting rocks. One pile especially intrigued him, so he filled his pockets. After finding Tonto, Jose mounted his burro and prepared to continue his journey.

"Instead of going on," he thought, "I'll go by Señor Lovin's and show him these rocks; he will think I have tried, even though I failed, and perhaps grubstake me once more."

Somewhat dejected,he paused before Lovin's store. Finally gathering enough courage, he entered hesitatingly and humbly offered his few samples. Henry Lovin looked at them, hefted them, and took them to the door for a better light.

"Jose, you have indeed found rich ore!"

The little Mexican looked up, startled!

"Your bill is paid with these few rocks, because they are worth far more than you owe me. Now take me to your mine!"

Thus was the discovery of the Gold Road Mine. Eventual gold production was about $7,300,000...a tribute to the wandering of a little burro and to Jose's laziness.

As some of Mohave County's most noteworthy gold producers, the Gold Road Mine and others in the district have continued to draw the attention of modern mineral exploration companies. Although some new resources of gold have been identified, there has been no recent production.

The Town Too Tough to Die

Tombstone, Cochise County.

Son of a 49er, Ed Schieffelin was a born prospector. Back in 1877, he traveled through the West looking for pay dirt. He joined up with a cavalry troop on its way through Arizona, then took a job protecting the Brunckow Mine from Indian attack.

One day, Ed was on guard duty when he was hailed by his boss, Al Sieber, who was a noted Indian scout himself.

"What're you doin' these days, Ed?"

"Prospecting, mostly," Ed answered. "Over yonder." He pointed eastward toward the hills.

"There?" Sieber scoffed. "All you'll find in them hills will be your tombstone."

Ed wasn't worried. When he set out for the desert hills — just as he expected—he made a strike.

After posting a claim, he took a few samples of ore up to Tucson for assaying. Finding no assayer in town, he decided to take the samples to his brother, Albert, who could tell the value of a rock as well as any professional.

Ed finally tracked Al down at the McCracken Mine in Mohave County. Richard Gird, the mine assayer, was asked to look at the samples. He was very impressed, and with the Schieffelin brothers headed back to the claim. It was rich with silver, as high as $2,000 a ton!

"Well, I got what I was looking for," Ed rejoiced. He remembered Sieber's warning and laughed."I guess this must be my tombstone."

Almost overnight, Tombstone became the biggest and rowdiest town in the entire West. The miners struck bonanza after bonanza—the Lucky Cuss, the Toughnut, the Emerald, the Contention, the Luck Sure and others whose names recall the exuberance of the day.

Between 1879 and 1891 the population grew from 1,000 to 7,000. And what a mixture! There were miners and muleskinners, madames and merchants, gunslingers, golddiggers and gamblers, prospectors, preachers and politicians from all over the West. And the entertainment district stayed open night after rollicking night.

Two out of three buildings were saloons or gambling halls. Lynchings were not uncommon, and the local cemetery encroached on the residential district so fast that a bigger area had to be found.

Typical of the daily activity was the feud between the Earps (whose shady activities were covered by their brother's position as U.S. Deputy Marshal) and the Clanton gang, who made sporadic forays into town from their hideout in the hills. The feud culminated with the famous shoot-out at the OK Corral.

Then Tombstone's heyday was cut off in mid-shout by dreaded underground water in the mines. Silver and copper ore remained beneath the ground, but pumping costs were prohibitive and mining work gradually sputtered to a halt.

Phelps Dodge bought some of the mines in 1914, and other companies continued extracting ore until 1950, but most of the $38 million in minerals had been found during the big ten-year boom. Now only the favorite spots of tourists—the Bird Cage Theater, the Crystal Palace Saloon, Boot Hill Cemetery and the OK Corral—survive in Tombstone..."the town too tough to die."

Wayne Winters, past publisher of the town's newspaper "The Epitaph" (every tombstone has to have an epitaph), is quick to remind us that Tombstone and her mineral deposits are indeed too tough to die. Mining activity in the district has been continuous for more than 25 years. Some form of mining, be it continued prospecting, exploration drilling, underground mining, open pit mining, or reprocessing of old dumps and tailings, can be found in the silver rich hills of Tombstone.

A Lynx & Gold Fever

Walker Party, Lynx Creek, Hassayampa. Yavapai County.

"There's gold in those mountains," an Indian told Joseph Walker in 1862. His words ignited a passion that drove a party of thirty-four men across desert, forest and mountains, and into the most dangerous Apache territory in Arizona.

So anxious were the prospectors that they disregarded the usual packing precautions. They stored their flour and gunpowder in weatherproof boxes, but everything else was subjected to desert heat and mountain storms.

Within a month, the party's clothes were in shreds. They skinned deer and wore the hides; and after winning a skirmish with the Indians, they scavenged their enemies' garments. But before they could wear the garments they had to bury them in ant hills to get rid of the lice. To make matters worse, they had to cut the hides off their backs when they shrank after a rain.

Men driven crazy by the scorching desert heat were tied in their saddles by the others. Nothing was going to make the party turn back, not even the fierce Apaches who preyed on all travelers. In fact, Captain Walker decided to take the Apache leader, Mangas Coloradas, as a hostage. With the help of a U.S. Army detachment, the group not only captured but killed the chief.

His death put the Apaches on the warpath and made travel far more difficult. But the Walker Party relentlessly made its way up the Hassayampa River to a gold-filled creek where they camped. When Sam Miller, a member of the party, found a lynx lying in the water, he thought it was dead and leaned down to pick it up; it "came to life" and clawed Sam, who kicked the animal to death. Ever since, the creek has been known as Lynx Creek.

The party found gold everywhere they panned in the creek bed. Numerous placer mines were opened up and down the creek. Men worked in teams, one with a pick and the other with a gun, because Apaches were a constant threat.

With the Indian raids becoming more frequent, the group's biggest worry was that their source of supplies would be cut off. They decided on a daring venture. They buried all their belongings in a fresh mound, built a fire over it to conceal the location and made a ten-day journey across the desert to the villages of the friendly Pimas. There they bought food, sent out letters telling of their discovery and hightailed it back to camp. Everything was exactly as they had left it. Their cunning plan had paid off!

By 1886, a two-mile stretch along the creek held some 20 houses with at least 60 men panning gold. Mills and smelters were built, and schools and a post office proved that this was a community of people who planned to stay.

They did...until the gold ran out. A few prospectors returned to Lynx Creek during the 1930s, but found little gold of real consequence. Although the creek yielded more than $2 million, the largest amount of any stream bed in the state, most of it was found during the first years of activity when the gold-choked river attracted an army of men who had traveled hundreds of miles for a dream such as this.

Minerals in a mining district seldom, if ever, really run out; they just become too rare to mine at a profit under current conditions of technology, markets, costs, and politics. A portion of Lynx Creek has been set aside for recreational gold panning. Exact directions to the Lynx Lake Gold Panning Area are included in the Arizona Department of Mines and Mineral Resources publication titled "Gold Panning in Arizona."

Mules & Copper Ride the Rails

Morenci, Clifton, Metcalf, Greenlee County.

The Apaches had struck during the night and made off with the best horses in the entire ranching area around Silver City, New Mexico. Led by Captain Chase of the United States Army, a band of cavalrymen took off in hot pursuit.

It was while they were scouting for signs of the Apaches' trail that Jim and Bob Metcalf first laid eyes on the site that was to become the largest copper mine in Arizona. They and several of their fellow soldier-prospectors staked claims, but little work was done at the time.

In 1872, encouraged by a truce with the Indians that made the Clifton/Morenci area safer, the Metcalfs returned to their claims and quickly realized that an embryo copper mine couldn't be developed on a soldier's pay.

Most of their needed capital came from the Lezinsky brothers of Las Cruces, who became deeply involved in the operation. The Lezinsky boys had an eye for business and constructed a smelter in Clifton. Before that, ore was shipped overseas to Wales for smelting.

The Lezinskys were resourceful. Their high-grade Longfellow Mine was five miles up the mountain from the new smelter. The brothers laid a narrow gauge railroad track to link the mine to the smelter.

Mules hauled the empty ore cars from the smelter back up the hill. And on the downgrade trip, the mules were passengers!

Some say the mules may have lodged complaints. Others say marauding Indians were too active. At any rate, the Lezinskys brought in a small locomotive, probably the first in Arizona. Today, an engine identical to the original locomotive is on display in downtown Clifton.

And so began what eventually became the Arizona Copper Company.

At about the same time, William Church arrived in Morenci, obtained options on four mining claims and organized the Detroit Copper Mining Company. His men worked side by side with those from the Arizona Copper Company, sharing tools and information freely.

In 1881, Church sold a major interest in his mines to Phelps, Dodge & Co., a New York import/export firm. He had entered their offices unannounced and coolly requested $50,000 to develop the Arizona mines. Intrigued by his confidence and sincerity, the Phelps Dodge partners decided to buy—and thereby laid the foundations of a multi-million dollar mining corporation.

The Detroit Copper Mining Company and the Arizona Copper Company continued to work closely together. They built a railroad and a 100-ton copper concentrator. They took big risks, and made big profits. By 1902 their combined annual output was over 24,000 tons of copper.

The Morenci and Metcalf mines produced steadily for the next several years, except for occasional setbacks caused by labor disputes and the fluctuating market value of copper. When World War I ended and the military demand for copper ceased, the Arizona Copper Company found itself on the edge of bankruptcy and sold out to Phelps Dodge. Now the entire area was managed by a single corporation.

Phelps Dodge made a bold decision to convert to an open pit mining system at Morenci. Production increased dramatically and has grown ever since. The Morenci pit is now the second largest copper mine in the nation.

If the ghosts of Bob Metcalf and the Lezinsky brothers ever return to the scene, they would be completely astonished by Phelps Dodge's great open pit operations, where one of the world's largest copper ore concentrators produces more copper in a single shift than the men of their era could have mined in a lifetime!

Phelps Dodge's Morenci copper mine and mill complex continues to be the largest copper producing complex in Arizona. Over the last few years it has become the largest in the United States.

Return of a Lost Soldier

Silver King and Silver Queen Mines. Superior, Pinal County.

Apache Leap is a high, red-streaked cliff just off Highway 60 near present-day Superior, Arizona.

In the early 1870s, a heliograph operator on Picket Post Mountain, nearby, watched the retreat of an Apache raiding party and discovered the long-sought base from which the Indians operated.

The seventy-five Indians, trapped by troops from nearby Camp Pinal, were unwilling to surrender and hurled themselves to death over the edge of the cliff now named for that desperate act.

Later, under those rugged cliffs of Apache Leap, Magma

Copper Company developed a new deposit of copper ore after the hot, deep shafts and deadly long drifts of the Silver King and Silver Queen mines ran out of silver. Ultimately Magma developed a rich ore body of ten million tons, averaging 5.8 percent copper!

That was the conclusion of a romantic story associated with the very beginning of mining at Superior.

In 1870 Camp Pinal, near Superior, was garrisoned by 400 cavalrymen in a canyon of sheer rock walls, winding down a narrow gorge carved by the flood waters of restless Queen Creek. Camp activities included pursuing Apaches, doing chores and building roads.

Trooper Sullivan was returning to camp from a few days of the latter enterprise when he sat down to rest for a while. As men are prone to do in mining country, he idly examined the loose pieces of rock at his feet. Among them were some heavy black fragments that would flatten but not shatter, and those aroused his interest.

Sullivan took them to his friend Charles G. Mason to ask if they had any value. Mason recognized them as specimens of metallic silver and suggested that Sullivan go back for a few more specimens and stake a claim while he was doing it.

Sullivan mysteriously disappeared. His friend suspected an Apache attack because the Indians had been particularly violent in that neighborhood.

After a long search, the original outcropping of black rocks was rediscovered and thus, in 1875, the Silver King Mine became another of the rich bonanzas in Arizona's history, paying handsome dividends to its early owners.

This discovery of the rich Silver King Mine first attracted attention to the district. Subsequent opening of the Silver Queen Mine, never as productive, increased the influx of fortune seekers.

A town called Pinal City grew up on Queen Creek, at the foot of Picket Post Mountain, as a major milling site. The silver finally played out, but fortunately large underlying deposits of copper were encountered.

Years after the discovery of the Silver King, an aged man wandered into camp and announced that he was Trooper Sullivan, original discoverer of the site. He had gone to California and in spite of a great desire to return had not been able to save money for the trip.

When he heard of the great success of the Silver King, he longed to return to see it with his own eyes before he died. It was a plausible story. The owners of the Silver King believed he was telling the truth.

So, Trooper Sullivan received care for the few remaining years of his life from having picked up a handful of heavy black rocks almost fifty years earlier!

Although closed down during a large part of the 1980s, Magma Copper Company's Magma Mine at Superior is again producing copper from the deepest copper mine in Arizona. Portions of the underground mine are so hot that workers are required to take 15-minute breaks after 15 minutes of work.

Pauline Was No Lady

Colorado placer mines, Weaver, Rich Hill. Yavapai County.

His Indian name was Quah-a-ha-na, which means "peace-maker" or "good talker." He was the son of an English father and Cherokee mother.

The Arizona Apaches liked Quah-a-ha-na, and in his early days Pauline Weaver was known across the territory as a friend of the red man.

He was tall and broad-chested with the shaggy beard and long locks of a "mountain man." Deep penetrating eyes stabbed out from his sharp bony features and his strong resonant voice was like the growling of thunder. He was one hundred percent man, despite the name "Pauline."

Born on the Tennessee frontier in 1800, he came westward, seeking beaver pelts along the Gila River.

During one of his long treks into southern Arizona, Weaver took shelter in the Casa Grande ruins and carved his name there, with the date, 1832, for all to see.

Arizona then was a place open for exploration and a prospector's pick at any blow might reveal a rich silver or gold vein.

So when not hunting, trapping or guiding, Pauline Weaver tried his hand at prospecting. A.H. Peeples and Weaver organized an exploration party. Around 1863 they left the Colorado River at a point north of present-day Ehrenberg and followed faint traces of washes eastward into the low mountains north of what today is the town of Congress.

One morning four Mexican wranglers were dispatched to find several horses that had wandered away from camp. They returned with not only the missing stock, but with some interesting rocks they had picked up on the summit of a nearby hill.

Weaver's curiosity was aroused, and he led the group back to the hill, where he used his hunting knife to pry loose a few good-sized nuggets. He carried one to Yuma and showed it to Jose Redondo, who declared it was gold!

Others abandoned the Colorado River to join the Peeples-Weaver party in the hills. At the original location, nuggets were strewn by nature in such a careless and opulent way that the spot was dubbed Rich Hill. It became the richest single placer discovery ever made in Arizona. Additional gold was found at Antelope Hill, nearby, where again much of it could be pried loose with a hunting knife.

The influx of miners around the discoveries turned the "Colorado placers" into towns. One of them was La Paz, a city now dead and forgotten in its own century. During its fierce and thriving years, La Paz missed becoming the capital of the territory by one vote!

Pauline Weaver is still identified as one of the discoverers of those placers east of La Paz. But, even as the city vanished, so has the gold.

With higher gold prices in the 1970s, 80s and 90s, places like La Paz, Weaver Creek, Antelope Creek, and Rich Hill became "alive" with miners just as they had been in the 1840s, 50s and 60s. The "modern gold rush" was different only in that the miner's burro had been replaced by a 4x4 truck, the handmade rocker by a portable gasoline powered suction dredge, the miner's metal pan by the latest design injection-molded polyvinylchloride equivalent, and the bed roll, cook pot, and canvas tent by the latest Coleman® camping equipment or fully contained motor home. Gold is still found, some recovered, but seldom enough to cover expenses, and naive investors are still taken.

Bill Hope & The Silver Sphere

Globe, Miami, Inspiration, Gila County.

The round of drinks that Bill Hope ordered for the soldiers proved to be a great investment. As they sat sipping their whiskey, one of the soldiers mentioned that he'd found some interesting rocks someplace in Apache territory.

His samples looked like silver ore to Hope. "Where'd you find these?" he asked casually.

"Oh...while we were chasing Indians. It was near a spring... by some wild cherry trees...." The soldier, who could draw better than he could prospect, made a map for Hope and went on his way.

Bill found himself a partner, Chilson, and they headed for the Richmond Basin in Arizona...the same area scouted by Ben Regan in 1874. Chilson explored the lower edge and reported there was nothing there, then returned to California. Hope continued prospecting on the upper edge, and uncovered a globe-shaped boulder of almost pure silver, with surface scars said to resemble the continents of the earth. It came from a great rich vein of silver.

A rush ensued, complete with the hasty construction of stores, saloons and cabins. In 1876, the miners indulged in a moment of braggadocio and gave their camp the impressive name of Globe.

Meanwhile, Hope's old partner Chilson returned and immediately "discovered" silver in the same spot he had earlier pronounced worthless. Hope was understandably suspicious, but Chilson insisted it was purely coincidental that he had struck silver after the partnership was ended. The credit for his discovery, he maintained stoutly, was all due to a divining rod! For some reason, no other divining rod ever discovered another ore body.

The bustling camp was right on the edge of Apache territory, and had to be officially ceded to the white men by the government. Before the papers came through, the camp faced a problem with liquor sales, prohibited on an Indian reservation. One enterprising fellow sold potatoes for a dollar apiece...and with every sale, threw in a free pint of whiskey as a goodwill gesture. His business flourished until he finally exhausted the area's entire supply of potatoes.

Although silver was the basis of Globe's first boom, copper later became its mainstay. After the surrender of Apache Chief Geronimo in 1886, it was safe to develop the property, and the Lewisohn brothers of New York took over the Old Dominion Mine at Globe. They built a smelter and a railroad, and for 25 years the copper mine was one of the greatest in the world.

By the time the Old Dominion gave out, emphasis had shifted to nearby Miami. There the Miami Copper Company has operated the Miami, Pinto Valley, Castle Dome and Copper Cities mines. The Inspiration Consolidated Copper Company has worked in the area as well, with the Inspiration, Live Oak, Thornton, Christmas and Oxhide mines. From 1916 well into the 1930s, copper production from the Globe/Miami/Inspiration area surpassed that of anyplace else in the state!

Today they're still producing copper at the various mines in the area. Although the mining methods have changed, the productive spirit is as strong now as it was in the days when men thought nothing of paying a full dollar for a potato and a token of goodwill.

Magma Copper Company now owns and operates those properties previously held by the Miami Copper Company, and Cyprus Copper Company now owns and operates those of the Inspiration Consolidated Copper Company. Copper now leaves the District as pure copper in rolls of 5/8 inch diameter rod destined for electric and communications cable factories in Arizona.

Grand Canyon Trails

Copper Canyon, Havasu and Orphan Mines. Coconino County.

For more than a million years, the Colorado River has been cutting and mining the awesome Grand Canyon—a vast gorge across northern Arizona. The river has uncovered deposits of metallic outcroppings and ore.

Many of today's trails into the canyon follow pathways created by early prospectors who searched for mineral wealth.

Countless mining claims were staked in the Grand Canyon. The stories of a few indicate the fever that existed.

The Canyon Copper Mine is in the very depths of the gorge, below Grand View Point. It was operated early in the century by two men—Barry and Cameron—who wore down a mile-long trail from the rim to the canyon floor.

The ore had to be packed up that long, winding path on mules. John H. Page, who later became a prominent Phoenix land attorney, made more than 100 such trips, herding the pack trains up and down the trail.

Later, the mine was bought by William Randolph Hearst interests and eventually became part of the Grand Canyon National Park. In fact, the trail is a major part of today's Grand View Trail.

The Havasu Mine was located deep in Havasu Canyon, a branch of the Grand Canyon. Mineralization was discovered in the 1860s on the sheer canyon wall. Prospectors could see outcroppings, but couldn't get to them. Finally, ingenuity triumphed. They got ore samples by shattering the face of the canyon with slugs from their rifles! Then they fashioned tall ladders to reach the prize.

At the turn of the century, W.I. Johnson of Prescott bought the lead and zinc deposits of the Havasu Mine. Power was a problem, and he conceived the idea of harnessing lovely Bridal Veil Falls for energy. Fortunately, this plan was never realized and the property eventually became part of the Grand Canyon National Park.

One of Teddy Roosevelt's Rough Riders, John Hogan, prospected in the canyon after the Spanish-American War. He found copper traces less than a mile west of Grand Canyon Village, 1,500 feet below the south rim. He filed his claim and called it the Orphan Mine. (The area had not yet become a national park.)

Hogan received a patent, some say, because he had influence at the White House due to his war record. He hacked out a trail to the mine. Next came a succession of ladders to the rim. The copper deposit turned out to be poor, but Hogan hung onto his patent. Finally operations ceased.

The years slipped by until 1951 when tragedy brought Hogan's mine back into the news. Two hotel employees, a young man and a young woman, were sitting on the edge of the canyon when the boy slipped, plummeting to a ledge 1,500 feet below. Park rangers managed to reach the body, using parts of Bright Angel Trail and the old Hogan Trail.

During this sad mission, they looked over the Hogan workings and were struck by an unusual yellow incrustation at the mine site. It turned out to be rich uranium ore.

The Western Gold and Mining Company eventually acquired the patent and built an aerial tramway from the diggings to the rim to transport the ore.

The Arizona Gold and Uranium Company currently owns the Orphan Mine, but shut down operations in the mid-1970s.

The property at the site of the Orphan Mine became part of the Grand Canyon National Park in the early 1990s. The headframe still stands as a reminder of the influence prospectors and miners had on early developments of the Grand Canyon area. A debate continues on whether the headframe should be allowed to stand or be torn down.

Tucson's Phantom Mines

Geophysical exploration, the cluster of mines. Pima County

The colorful, bearded prospector, with his faithful burro, has a special niche in Arizona history. Rare as a cigar store Indian today, he has left a vital heritage to the modern mining explorer.

With his rule-of-thumb geology he effectively mapped the areas of mineralization to provide the launching pads for the engineer of the present. The elusive quarry today is the deep-lying ore body. It might be called a "phantom" mine.

For centuries, conquistadors and prospectors tramped within two hundred feet of many ore deposits, but never found them—because the two hundred feet was straight down beneath the surface and the phantoms were well hidden.

Today, modern technicians—geologists, geophysicists, geochemists, geobotanists—call on every recent scientific development to give an area the "third degree."

Scientists, for instance, may trail instruments behind an airplane, watching needles to tell them there's "something different" down deep under the ground. They may lay a grid of wires over the area and take the same readings.

They charge the ground with an electric current to fathom the conductivity of what is below. They may drill holes and set off explosives to read on a seismograph the difference in the "thud" that results. The geobotanist even tests the plants growing in the area for traces of copper.

Painstakingly and exhaustively the process goes on. Then countless test holes are drilled and the samples are carefully assayed. More often than not, the results are negative. But sometimes a major ore body is located.

The area south of Tucson has become a fabulous treasure trove with a string of major copper deposits. Early discoveries go back over a hundred years, but the amount of mining was moderate compared to today's massive operations.

What happened? Simply, in the late 1940s geophysicists and other scientists became prospectors. The Pima Mine was the first discovery by the new methods.

Not every mine in the Tucson cluster is the result of geophysical search, but the big producers are. The roster of companies and their mines is impressive: Anamax Mining's Twin Buttes, Eisenhower's Palo Verde, American Smelting and Refining's San Xavier and Mission Units, Cyprus' Pima, Duval's Esperanza and Sierrita.

History's bearded old prospector would be speechless, to say the least!

Cyprus Copper Company now operates the Twin Buttes Mine and the Sierrita Mine, and Asarco, previously known as American Smelting and Refining Company, operates the Mission Mine and the San Xavier Mine. The others have been combined into these operations.

The Arizona Mining & Mineral Museum...

Arizona's Mining and Mineral Museum attracts thousands of visitors each year—students, area residents, visitors from out of state and from around the world, hobbyists, collectors and prospectors. It contains more than 15,000 specimens including azurite, the brilliant blue copper oxide, obsidian, turquoise and malachite.

The Museum is the official repository for the identification, cataloging and display of specimens, ores, gemstones and lapidary material found in Arizona. Visitors are invited to attend the Museum gift shop where mineral specimens, jewelry made from minerals, publications relating to mining and minerals and a wide range of other items are for sale. Located in Phoenix at 1502 W. Washington, the Musuem is open Monday–Friday from 8 AM to 5 PM and Saturday from 1 to 5 PM. Admission is free.

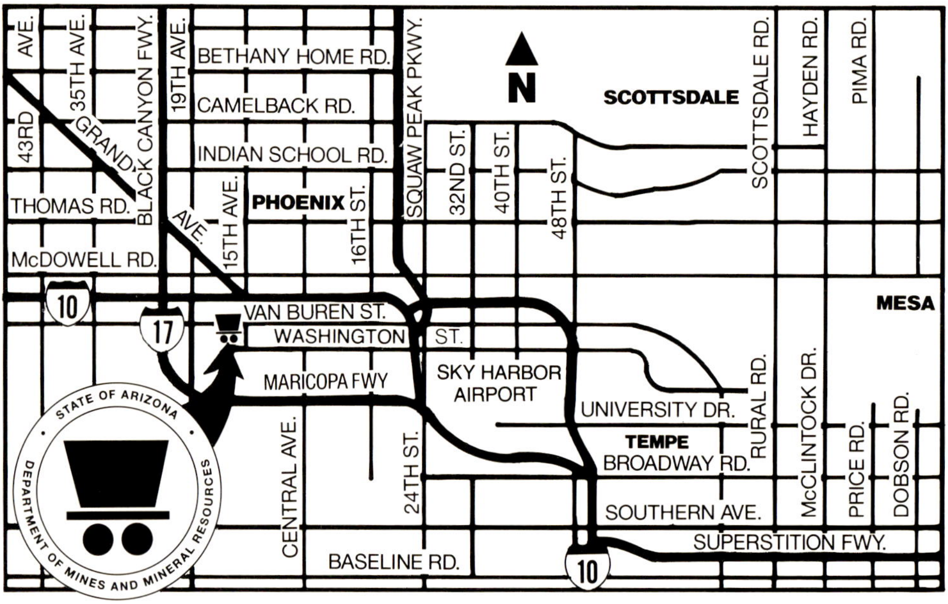

Index